Plants as Food

A & C BLACK • LONDON

Plants as Food

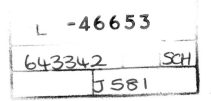
© Blake Publishing Pty Ltd 2002
Additional Material © A & C Black Publishers Ltd 2003

First published 2002 in Australia by Blake Education Pty Ltd

This edition published 2003 in the United Kingdom by
A&C Black Publishers Ltd, 37 Soho Square, London W1D 3QZ
www.acblack.com

ISBN 0-7136-6597-1

A CIP record for this book is available from the British Library.

Written by Paul McEvoy
Science Consultant: Dr Will Edwards, School of Tropical Biology,
James Cook University
Design and layout by The Modern Art Production Group
Photos by Photodisc, Stockbyte, John Foxx, Corbis, Imagin,
Artville and Corel

UK Series Consultant: Julie Garnett

Printed in Hong Kong by Wing King Tong Co Ltd

A & C Black uses paper produced with elemental chlorine-free pulp,
harvested from managed sustainable forests.

Plants Feed the World

Plants are the only living things that can make their own food.

Every day the sun's **energy** shines down on the earth as light. Plants use this energy to make their own food. Each leaf on a plant is a food factory that makes simple sugars. These sugars feed the plant as it grows. Plants then become food for other living things.

Plants are the main food for many animals. Animals that eat plants are **herbivores**. Hoofed animals, such as cows, zebras and deer, eat grass. Monkeys, bats and other small mammals eat leaves and fruit.

Carnivores are meat-eaters. However without plants even carnivores could not survive. Carnivores eat animals that feed on plants, so all living things depend on plants.

Cows are herbivores.
They eat grass.

Zebras eat grass.

Lions are carnivores.
They eat meat.

5

The Food Chain

Food chains show how plants and animals are linked together by what they eat. Most food chains begin with plants.

Plants and animals are connected to one another like the links in a chain. At one end is the carnivore; at the other end are plants. Lions eat zebras. Zebras eat grass. This is known as a food chain.

Each habitat has many different food chains. In oceans, deserts, forests and grasslands, plants and animals are linked together. At ground level in a forest, rabbits eat grass. Foxes and wolves then eat some of the rabbits. In the trees, squirrels eat acorns and then get eaten by owls.

Some food chains are longer. Grasshoppers eat plants. Frogs, lizards and mice eat grasshoppers. Snakes prey on mice and frogs. Birds of prey, such as hawks, hunt and eat snakes.

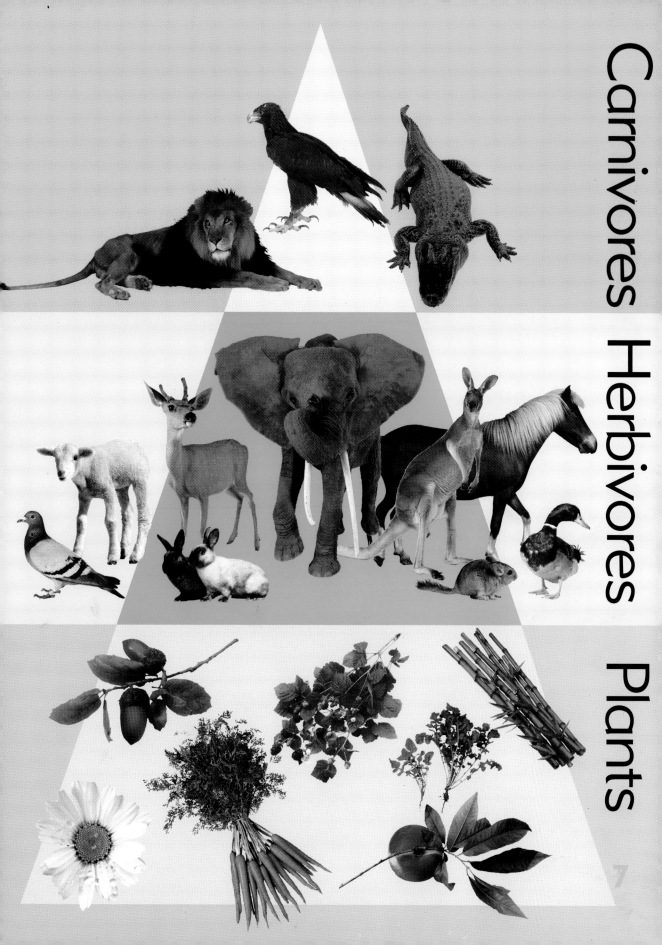

What Do We Eat?

People can eat many plants, but not all of them. You need to know which part of a plant you can eat. It can be the fruit, nuts, roots, leaves, stems, seeds or flowers.

Fruits

Fruits contain seeds covered by a tasty treat. Oranges, grapes, melons and berries are all fruits that people eat. Many animals eat fruits too.

orange trees

8

Roots

Many plants store their food in roots and **tubers** under the ground. People eat roots like carrots, and tubers like potatoes. Porcupines dig up and eat roots and tubers.

Nuts

Nuts have a hard shell around a seed. People eat nuts such as cashews, walnuts and almonds. Some animals, such as squirrels, collect and store nuts for winter.

lettuce

Leaves and stems

People eat the leaves and stems of lettuce, spinach and celery. Many animals like leaves and stems. Koalas feed only on eucalyptus leaves. Pandas eat a lot of bamboo to live and grow.

celery

Seeds

Plant seeds are the most important part of the plant for people. Most people in the world eat seeds every day. Rice, wheat and beans are all plant seeds. Many animals, including pigeons, parrots and mice, eat seeds.

rice

a wheat field

10

Flowers

People eat the flowers of broccoli and cauliflower plants. The sweet **nectar** in other flowers feeds birds, bats and bees. From this nectar, bees make honey to feed their young. We eat honey too.

broccoli

cauliflower

bee collecting nectar

From Seeds to Food

Vegetables take months to grow. Fruit and nut trees can take years to grow, but the sprouts of some seeds can be eaten after just a few days.

From a seed to your table, how long does it take?

What you need:

- a packet of alfalfa seeds
- a plastic container with a lid
- a tablespoon
- water and sunlight

What to do:

1. Place 2 tablespoonfuls of seeds in the container.

2. Cover the seeds with warm water. Place the lid on top and leave overnight.

3. Drain off the water and leave the container in a warm, well-lit spot.

4. Rinse the seeds under cold water twice a day. Drain well each time.

After 6 days, the sprouted seeds should be ready. Wash them before you eat them.

Day 1

Day 2

Day 3

Day 4

Day 5

Day 6

Grains

Wheat, rice and corn are the seeds, or grains, from grass plants. Grains are the main source of food for most people.

Grains are the seeds of grass plants. Grasses have thin stalks that grow into a head of seeds. Wheat, rice, corn and oats are all grains.

Wheat is ground into flour to make bread and pasta. Puffed wheat can be made into breakfast cereals.

Rice is the main food for half of the world's population. Rice is often cooked and eaten whole. It can be ground into flour to make noodles.

Corn or maize is ground and made into tortillas, cornbread or cereal. Many farm animals, such as cows and chickens, are fed corn.

These young rice plants are about to be planted in a rice paddy.

Wheat grows in large fields.

This bread is made from wheat flour.

Fruit

Some plants produce seeds inside fleshy fruits. Fruit can grow on a tree, bush or vine.

Oranges, lemons and limes are all citrus fruits that grow on trees. Citrus fruits need long, hot summers to grow well.

Stone fruits, fruits with stones in them, also grow on trees. They have one hard seed covered with soft flesh. Peaches, plums, cherries and apricots are stone fruits.

Many fruits are quite small. Strawberries, raspberries and blackberries are all small fruits with lots of seeds. They grow on small plants or bushes in cool areas.

Apples and pears grow on trees in cool areas. They both have a core with small seeds inside. Some apples are grown to make juice or cider to drink.

Melons grow on vines that trail along the ground. Grapes grow on vines that grow along fences.

Strawberries grow
on low plants.

Cherries grow
on trees.

Grapes grow on vines.

17

Vegetables

People eat hundreds of different vegetables. They are an important food for people around the world.

We eat the stems, leaves and flowers of some plants. Lettuce and spinach are leafy vegetables. The stems of celery and asparagus are good to eat. Cauliflower and broccoli are edible flowers.

Pumpkins and tomatoes are vegetables that contain seeds. We also eat the seeds and **pods** of peas and beans. Scientists call them fruit because they are the part of the plant that carries the seed. When we cook and eat them, we know them as vegetables.

Some vegetables grow under the ground. We eat carrot roots and potato tubers. We also eat the **bulb** of the onion plant.

The seeds or fruits of different plants are crushed to make vegetable oils. We get oil from olives, sunflowers and walnuts.

These ripe pumpkins are growing in a pumpkin patch.

People eat sunflower seeds, but most sunflowers are grown for their oil.

Cucumbers grow on vines.

19

Foods for Fun

Food gives us energy and keeps us healthy. Food can also be made sweet or spicy just for fun.

Many animals have a "sweet tooth". Birds and bees drink sweet nectar from flowers, and bears eat honey. People eat sugar made from the dried juice of sugar cane.

Herbs and spices are used in cooking. Herbs such as basil and parsley are used as seasoning. Garlic adds flavour, and chillies are hot and spicy.

Chocolate, vanilla and cinnamon are also plant flavours. Chocolate is made from seeds. Vanilla is made from seed pods, and cinnamon is ground from the dried bark of a tree.

Many drinks are made using plants. Coffee beans and tea leaves both come from plants. Lemonade is made from the juice of lemons.

Sugar comes from sugar cane.

Chocolate is made from cacao seeds.

Vanilla flavouring comes from seed pods.

21

Plants as Food

seeds	
roots, tubers and bulbs	
leaves	
flowers	
fruit	
nuts	

Glossary

bulb   the underground part from which some plants grow

carnivores   animals that eat meat

energy   power to do, make, or grow

herbivores   animals that eat plants

nectar   a sweet liquid that flowers make

pod   a seed container that grows on plants

tuber   a fleshy growth under the ground, such as a potato

Index